Nucs

By Dave Corbett

Northern Bee Books

Nucs

ISBN 978-1-914934-60-5

Published by Northern Bee Books, 2023
Scout Bottom Farm
Mytholmroyd
Hebden Bridge HX7 5JS (UK)
www.northernbeebooks.co.uk

DM Design and Print

CONTENTS

CHAPTER 1

WHAT IS A NUC?

A nuc, or nucleus colony, is simply a small colony of bees that have been created by a beekeeper to use for a range of purposes. This is where the simplicity ends! To go beyond that definition, it is important to explore some of the reasons that beekeepers want to create smaller colonies – after all much of the season is spent trying to produce and maintain the largest colonies possible to maximise their nectar gathering potential and the subsequent honey crop!

Beginning with the format that most beekeepers are familiar with, a nuc can be a half size colony. To this end the BBKA have attempted to produce guidance for what it should be. In 2009, they created the leaflet BBKA Standard and Guidance Notes for Nuclei – as a replacement for the withdrawn British Standard BS:1372 (Bees, Colonies and Nuclei) that had been in circulation since 1947. Broadly speaking this was a good basis, containing the following provisions -

▶ A well-balanced colony of between 3 and 5 brood BS brood frames.

▶ Bees, food, brood and a queen.

▶ No active queen cells.

▶ No more than 15% drone brood.

▶ Not less than 30% sealed brood.

▶ Queen marked in standard colour (may be clipped).

▶ Healthy (eg. Free of disease).

However, this definition does have a few problems. Primarily, it is only really describing what might be expected in a nuc for sale. This is useful to those purchasing nucs in the early years of their beekeeping, but does not really offer much to those creating nuclei for other purposes. Additionally, the description is presumably based on 5 frame nuc boxes, presumably the wooden type used by many before the ploy nuc took hold. Every poly nuc that I have encountered is 6 frame, as are many of the correx nuc boxes that many nuc producers now sell bees in.

Another purpose of a nuc, not covered by the above description, is to enable to beekeeper to make increase. This is the one of the main focuses of this booklet. In the UK, there is always a shortage of nuclei available in the early portion of the year. For the hobbyist beekeeper, this can cause some difficulties and expense. The dearth of nuclei available in the spring means that the cost of a nuc at this time is always higher, if there is one available at all – commercial producers often sell out early and fellow smaller scale beekeepers will rarely produce enough to match the demand. The solution? To produce your own – they act as an insurance policy to cover any winter losses that you may suffer and can be used to expand your operation in the new season – quickly growing into full sized colonies in the spring. However, the creation of a nuc for increase may be very different from that of one that is ready for sale.

Nucs can be handy to have for demonstration purposes. The sight of an observation hive on a stall or display is fascinating for the general public and will often attract quite a crowd. For most people it is a novel experience to be able to see honeybees up close. Alternatively, a nuc can be a great introduction for beginner beekeepers – not too large to be intimidating, but enabling the new beekeeper to develop confidence in inspections.

Swarm control is another useful way in which a nuc box can be utilised. For many beekeepers, May can be a time when equipment is at a pinch point – swarm control using techniques such as the Pagden method can require significant additional equipment. However, removing a queen and a frame of brood or two into a nuc box is often simple and

Mating nucs are a different type of nuc entirely. There are the mating nuclei that use full sizes frames, often containing 2 or 3 to provide the resources to sustain the bees whilst the queen mates. These are useful to the beekeeper as they are relatively straightforward to create and allow for easy expansion. There are also mini nucs, are miniature hives that are stocked with minimal resources to enable a queen to make her mating flights and demonstrate laying but very little else. There are an enormous variety of brands, such as Apidea and Kieler among others. These use frames which are much smaller in dimension that standard frames, and usually have internal feeders in which syrup or fondant is given as there are not enough bees to support itself. These are commonly used because they avoid the need to deplete production colonies in order to get new queens mated.

CHAPTER 2

CREATING SPLITS

Most beekeepers will be familiar with the process of splitting colonies, having almost invariably been taught a method if they attended beginners training or having researched and practiced one (or more) if they didn't. Essentially, creating nuclei is a very similar process.

Before examining the variety of methods that could be used, there are a few golden rules in terms of creating nuclei that will avoid meeting difficulties further down the line.

Using Healthy Colonies

This should really go without saying, but using healthy colonies is of utmost importance. Generally, healthy colonies are likely to be the ones that are well established by the time that you are ready to make splits. As always, vigilance for the signs of European Foul Brood and American Foulbrood is essential. However, perhaps less obvious might be the presence of less serious disease. Chalkbrood whilst not significant and usually resolved alone would most likely preclude the use of eggs/larvae from – as there is a genetic predisposition toward the management of fungus by colonies. Equally, a smaller nuc will temporarily be under more stress potentially leading to the proliferation of the condition. In creating some of the types of nuclei that will be described later, it is a good idea to use a similar selection criteria as if you were rearing queens.

The Right Amount of Bees

When transferring brood to create a nuc there are two considerations that are significant. The first is to ensure that there are enough bees to over any open brood. Larvae are particularly susceptible to temperature and so need enough nurse bees to maintain this. Sealed brood on the other hand is able to generate its own warmth and consequently requires less input from nurse bees. The best way to ensure that there are enough is to select the frame from the donor hive and then give it a firm tap. This will cause fliers to leave the comb, allowing for a more accurate assessment of whether there are enough to sustain any open brood. If there are not additional nurse bees can be added to the nuc by shaking them off another comb until a satisfactory number have been included.

Using Multiple Colonies

For those beekeepers with a number of colonies, a way to create nuclei whilst maintaining the strength of existing colonies is to use brood frames from multiple colonies. There is a caveat here – if bees are taken from 2 colonies the result will be fighting between them. It is necessary to use 3 or more to avoid this. However, for modest increase this does allow the beekeeper to cause minimal disruption to their production colonies.

Moving To A Different Apiary

An additional issue with the number of bees is if the nuc is going to remain in the same apiary (often the case for smaller scale beekeepers with a limited amount of space/sites to move nucs around). This is that if fliers are included within the creation of the nuc, they will quite quickly return to their original hive. This can leave what initially appeared to be a suitably strong nuc significantly depleted. Ideally, this is why it is better to move a nuc to a separate site in order to minimise any risk of this.

Food

As nuclei are a smaller colony of bees, they can at times struggle to provide for themselves. It is always a good idea when creating a nuc to add ample supplies of food, to at least see the nuc through to the following week. A frame of open nectar will provide for their needs and is easily accessible to use. Alternatively, many beekeepers remove excess stores after overwintering – to avoid contamination of their honey with any sugar syrup that may remain after autumn feeding. This is a good use of a resource if you are able to store such frames. Of course, another method is to provide feed in the form of sugar syrup – but this does tend to require monitoring to ensure the nuc does not fill available brood space in the course of processing this. If the nuc itself is going to be producing it's own queen, as in methods such as the walkaway split, it is a good idea to provision them with a frame of pollen. This gives the resources to provide maximum nutrition to potential future queens.

Space To Expand

Nucs can grow at an extraordinary pace if the conditions are right – rapidly drawing out comb if foundation has been given or quickly filling that comb if it is already drawn. Whilst this is a good situation to be in, it also requires monitoring to ensure that the nuc does not become overly strong and potentially make swarming preparations.

Siting

The siting of nucs will largely depend on the apiary that they are placed in, but it is worth remembering that they do not have the ability to regulate their temperature as easily as a full sized colony. As such it is a good idea to try to place them out of direct sunlight to avoid the risk of overheating. Poly ncs are better that wooden nucs in this regard as their temperature tends to remain more stable due to their enhanced insulation.

Types of Split

Walkway Splits

This is the simplest method for all beekeepers to create nucs, requiring the least input and is therefore an option for most looking to create new colonies. With that said, there are various nuances to using this method and it is important to understand the processes that are happening to increase the level of success.

At the most basic level, a colony is divided into two (or more) parts – one retaining the original queen and the other containing the necessary elements to raise its own queen. In fact, this allows the beekeeper the possibility to not even be certain which half of the split contains the queen as long as both have the resources to raise a new one. The name walkaway comes from the leaving the colonies alone for around a month, in which time if things are successful a new queen should have been raised and be beginning to come into lay.

However, although this initially seems straightforward this method still seems to cause a number of questions – most of which can be resolved but as it will become apparent the name walkaway is a bit of a misnomer. The first issue is when to split. There are several elements to this. The first is when to make the split – the original colony will need to be at the point of being very strong. Personally, I like to encourage colonies to build up to double brood if I am going to use this type of split, and then create two of three nucs from one half of the colony leaving the donor colony with the queen to allow that to continue to be productive in terms of honey production. It would be unusual for this to be the case much in advance of the spring flows, and depending on weather conditions may not be until May. Even if splitting a colony that is using a single brood box in two, it is unlikely to be much earlier than this because of the second factor. This is the presence of drones for virgin queens to mate with. There is simply no point attempting this method too early (or late) in the season if there are not going to be enough drones available. Whilst you cannot be expected to know what is happening in the colonies of other beekeepers in the nearby locality, what is happening in your own hives is often a good indication. Are your colonies raising drone brood? Have you seen emerged drone on the combs? If the answer to these questions are yes, then this is the earliest you would consider this type of split. Although drones take 14 days from emergence to mature, if a split is made once drones are seen in the hive – those drones would be mature by the time a queen is ready to go on a mating flight. It is highly likely that a similar stage of colony development is going to be happening for the other beekeepers nearby.

Once you have decided that it is the time to split, the third consideration is the cells that are produced themselves. When making up the split, you will need to provide the part which is going to raise its own queen with very young larvae with which to do so. For most, this will mean using frames containing eggs, larvae and brood. It is the larvae which make this slightly more hands on than simply doing the split and walking away. One consideration is that bees left queenless but with the means to raise a new queen will produce emergency cells. These do not always result in the best queens as if the bees do not have larvae at the right age they can't provide optimum nutrition. Consequently, these queens tend to be smaller and less prolific. This can be remedied by marking a frame containing eggs and then going back into the nuc to break down cells that are made on older larvae (on other frames) – but this is moving away from the idea of walking away! It is nonetheless a good idea to go back into the split and thin out the number of cells, leaving only one behind.

The final consideration is food – if the nuc is to be left for a month, it will need to be given enough supplies to feed itself during this time. These type of splits are usually able to manage during a honey flow if they are given some stores when created, but it is essential to consider whether they are able to feed themselves and possibly provide them with feed if conditions are adverse. Again, potentially this moves away from the idea of simply splitting and walking away – but overall this is still an easy method that should be well within the capabilities of most beekeepers.

Using a Queen Cell or Adding A Queen

Another way in which a nuc can be easily made is to add a queen cell or queen to a nuc that has been made up for that purpose. This is slightly different to the use of mini mating nucs which is covered later, although several of the principles are similar.

The biggest difference between this and the walkaway split is that it is not desirable to provide the bees with the materials to raise their own queen. However, this does allow the nuc to be created in such a way that it is weaker – and therefore not involve as large a demand on resources. This is of particular benefit to the beekeeper that wishes to increase the number of colonies that they possess quickly, especially when using mated queens.

There are many sources of information on queen rearing, and as these are outside of the scope of this booklet they will not be covered. On this basis, the assumption will be that you have a queen cell that is at the 14 day stage – almost ready to emerge. This is important, as whilst a nuc could conceivably keep the cell warm enough to enable the queen to go through the earlier stages of development it is not really of a huge benefit to use a nuc to do this. To create the nuc, the key is to use well covered frames with lots of young bees. Ideally, the frames will be mostly emerging brood too – and certainly not any young larvae or eggs. This means the workers will have no alternative but to accept the soon to emerge queen. The cell itself can be protected – there are a variety of cell protectors/holders that can be used which will prevent damage to the cell and hold it in place. Alternatively, wrapping the sides of the cell in tin foil will serve to protect it. Place the cell in between two frames of brood and then the most important skill here is to exercise patience. The queen should emerge in two days time if the timings are correct. She is then likely to be ready to take mating flights in around five days. This is dependent on the weather – if conditions are good she may fly marginally earlier, but be prepared that weather conditions may also delay her mating flights. If she is successful in mating and returning to the nuc, she will then need time for the semen to migrate to her spermatheca before she can begin laying fertilised eggs. At the very soonest, it is likely to be two weeks from introducing the cell. Preferably, leave the nuc alone for three weeks – by which time she should be laying (although this can take longer). Earlier checks are not always productive, as young queens can take some time to settle into a proper laying pattern – often depositing multiple eggs in the base of cells, which can be mistaken for a problem by inexperienced beekeepers. Due to the length of time required, providing a nuc with adequate stores or feed is a good idea – particularly earlier in the season or during spells of dearth. However, this still allows this type of nuc to be made up of three frames quite effectively – especially during a flow (two frames of brood will often suffice in a good flow).

If introducing a mated queen, it is important she is caged in an introduction cage. If she has been purchased this is most likely how she will be received. If she has been home raised, then you will need to transfer her into an introduction cage yourself.

The need for young bees in the nuc is even more critical here – older bees are less accepting of a different queen. This is an additional use for a nuc made up using a mated queen – she can be introduced into a nuc with relative ease and then combined to a larger colony once she has built up. The next point to consider is the attendants that are in the cage with the queen. Whilst these are important for ensuring that a purchased queen arrives alive and well, they can be a potential source for her rejection. A home reared queen being added can easily be placed in an introduction cage without attendants – but a purchased queen will often come with five or six workers. These should be removed from the cage prior to introducing her to the nuc. This can be done within a clear plastic bag – enabling the beekeeper to contain all the bees and ensure the queen is the only remaining bee in the cage or easily recapture her if needed. The next step is to break the tab on the introduction cage. Some beekeepers choose to leave this for a day or so to further slow introduction, but this is not necessary as the fondant that the workers will eat through will serve to slow the queen's introduction to the colony. If the nuc to receive the queen has been made in advance, it will quickly realise that it is queenless and be in a prime state to receive the queen. Most cages can be hung easily between two frames. As an extra security measure, a toothpick can be used to prevent the cage slipping in between the frames.

CHAPTER 3:

NUCS IN THE AUTUMN

Nucs can be created throughout the active beekeeping season, and often quite late into the year – how late is a question that is often asked. With much of beekeeping in the UK, the local conditions will largely dictate this. In my local area, there is relatively little forage after late July until the ivy begins to flower. This has an impact on what is possible in creating splits as the amount of drones available for mating significantly reduces and perhaps the biggest challenge is the increasing challenge created by wasps.

Anecdotally, it has been suggested that even a small colony where everything is good will defend itself and I have managed to get queens mated in new splits into late August. But there is a significant failure rate – which can be very off putting, especially to those beginning their journey into raising their own nucs. If you are going to attempt this, it is really important to focus on particular areas to do everything within your ability to help the fledgling colony establish itself.

Queens

This is the most suitable time to use a mated queen – as it means there is less delay in the beginning of brood rearing and consequently the rest of the colony is more predisposed to defend itself. This can require a little planning if you are intending to purchase a queen as the availability of queens reduces as the season gets later. This is especially the case if you want to use a UK reared queen due to our climate and the comparatively short length of our season. If you are intent on allow the bees to raise their own queen, you will need to carefully work backwards with a calendar to establish when key moments will happen for the colony and how you will help them to be built up to a large enough size to successfully overwinter.

Food

Whilst a full colony will usually have stored enough honey to take care of their needs (assuming the beekeeper hasn't been too greedy in taking off a crop), a nuc that has recently been created is unlikely to be in this position and so will need additional resources to be given to it. There are several solutions to this. You may have some colonies that were honey bound in the spring – overfeeding in autumn, a strong ivy flow or a particularly frugal colony could be the reason – but the consequence is that you may have needed to remove some frames of stores. These are ideal to provide the nuc with food to facilitate their build up. The standard caveats apply – be certain they were from a disease free colony and that the stores are accessible (which mostly applies to ivy frames). The next alternative is to feed sugar syrup. This is easy in terms of availability – but again real care needs to be taken in terms of apiary hygiene regarding spillages needs to be taken to avoid attracting robbers or wasps. A final option is to provide fondant – it will give the colony the stores that it needs whilst avoiding attracting unwanted attention. Clearly, this needs to be weighed against the expense of this.

Entrances

As with full colonies, when wasps become more problematic it is a good idea to reduce the entrances. This is usually straight forward with wooden nuc boxes – mine came with rotating entrance blocks that can simply be fitted to achieve this. Polystyrene nucs are a little more difficult to reduce – the tunnel entrance leaves a relatively defensible entrance, but a lower number of bees means that these can still be at risk of attack. I tend to slide the opening disk round to half way and then use a drawing pin to hold it in place to further decrease the size of the entrance.

CHAPTER 4:

OVERWINTERING NUCS

One of the benefits of creating nucs is having the option of using these to ensure colonies go into winter with a young queen. Replacing older queens means that the colony is headed by a vigorous young queen who should be able to raise lots of winter bees helping with successful overwintering and a strong spring build up. Additionally, this should help to reduce the risk of swarming in the following season. If the queen has been allowed to develop on full sized frames, it is an easy task to remove the old queen and then use newspaper to unite the queenless colony to a queen right nuc. Once the colonies have been combined, they can either be overwintered on double brood or empty, older frames can be removed to bring the colony down to a single brood and facilitate comb change. If mini mating nucs have been used, it is a little trickier – but not significantly so. As before the old queen should be removed. The process for introducing a queen – using an introduction cage – can be followed allowing for the replacement to take place gradually.

This brings us on to the issue of the older queen. Review your notes, as she may well be worth keeping! If she hasn't made swarm preparations or she has been particularly calm in temperament, it would be a shame to waste those good genetics. A better option is to put her into a nuc – where the burden of overwintering and spring build up is reduced and therefore might enable her to form part of the following season's queen rearing programme.

Of course, the last and easiest option is to allow the queen to overwinter in the nuc that she is in. This is easily possible in 5/6 frame nucs, and achievable in three frame nucs. With mini mating nucs it is possible but will require a bit more forethought – larger mating nucs, such as the Mini Plus, especially when equipped with an extension can overwinter well, but smaller ones without the ability to provide a larger space will struggle.

Food vs. Brood

It is really important to check on nucs later in the season – to ensure that they have enough space for stores to enable them to overwinter. Whilst nucs, being small and with most people using polystyrene, do not require anywhere near the same amount of stores as a full sized colony in a wooden box, will still need to be able to provision itself with food for the winter. In a 6 frame nuc, I find two full frames before the ivy flow is usually enough to avoid having to feed at all over winter in my location. I prefer to overwinter colonies on single brood, so I often find that I will need to collapse colonies that have run on double brood down once I have extracted the honey they have produced at the end of July and early August. This often means that I have additional frames available to add to nucs and consequently can avoid needing to feed them at all. Those that are light as September arrives, I feed with 2:1 syrup.

However, adding food once the bees will no longer take down syrup is more challenging in several of the models of polystyrene nucs – often requiring less than satisfactory scenarios of removing parts of feeders or squeezing fondant into feeders (not ideal at times of the year when you want to minimise the length of time a colony is open for).

Earlier on in the season, additional brood can be used to bolster colonies, or if earlier on create further nucs. This is often beneficial as it avoids the need to weaken a honey producing colony in order to create a nuc and particularly useful to a small scale beekeeper who is trying to balance honey production with their desire to overwinter some additional nucs.

Varroa Treatment

Possibly the key to successful overwintering is varroa treatment as with larger colonies. The challenge with nucs is the dose to apply – most treatments are provided for full sized colonies and because of this some would be harmful if applied to a smaller colony because of the lack of volume in a nucleus colony. In the situation of needing to treat a nuc I have found the strip type treatments, such as Apivar and Apistan, to work well. Oxalic acid vapourisation seems to be effective if repeated over a brood cycle, but this is time consuming.

One fortunate factor in raising a queen in a nuc is the broodless period that they often go through prior to a new queen coming into lay. At this point, oxalic acid vapourisation can be a highly effective means of lowering the varroa count and getting a colony off to a good start. I haven't used oxalic acid dribble at this point, although I would presume that this would be equally effective. Again, a lot of this is about considering the time of year and the need to treat. A mid-winter treatment of oxalic acid is recommended – and seems to be a relatively easy way to ensure than the nuc is able to get off to a strong start in the spring.

Mouse Guards

Mouse guards can be a tricky balance between allowing a colony as much opportunity to gather pollen without the reduction in income through it being knocked off legs and the need to protect from mice. As many of my apiaries are in rural locations, it is not unusual for mice to attempt to enter colonies. A standard approach can be used with a wooden poly nuc – either by purchasing a mouseguard pre-cut to size or cutting down one for a hive. However, poly nucs do not have this option. The best solution that I have found is the reduction of the entrance as for wasps. Unfortunately, there are still times when mice have managed to eat away at entrances and get into hives – which can be particularly frustrating as this can result in the loss of an otherwise healthy nuc and often leaves damage to the nuc itself. Elevating the nucs further from the ground seems to help – but again is not a surefire method.

When Spring Arrives

Hopefully, all your colonies will come through winter healthy. One of the greatest disappointments I find in beekeeping is those colonies that don't. This is where raising your own nucs can be hugely beneficial. Having the ability to replace any losses is a massive advantage at the start of the year.

Spring can be a difficult period – nucs of all sizes can fill very rapidly, and I strongly suspect that a large number of early swarms are the result. Unfortunately, the weather doesn't always allow for inspections as early as we would like, so it is often a good idea to prioritise nucs. If you are looking to expand and plan to transfer the nuc to a full size hive, it is a good idea to use a dummy board to manage the number of frames as the colony expands.

A failed or failing queen can readily be replaced by using newspaper to combine colonies after removing the problem queen. If she is a drone layer, find her and despatch her, then combine the nuc to the colony. Alternatively, you can move the hive about 200 metres away and shake all the bees out. The nuc can then be installed on the original site and will get the benefit of an additional workforce as they return to the hive. Drawn frames from the original hive without drone brood can be used to save the colony the work in drawing combs. A similar measure can be used for laying workers.

CHAPTER 5:

THE NUC METHOD OF SWARM CONTROL

The nucleus method of swarm control is by far my favourite method for several reasons. The first being that is it much more practical for me to transport around nuc boxes at the height of swarming period than full sized hives. It simply makes my beekeeping more efficient at a point in the year when time is at a premium. The second reason is that I find that in preventing swarming in this way, the split that has been created holding the queen will often rebuild itself very rapidly – at times even returning to being a production colony before the end of the season. Finally, the ease of transporting nucs means that this allows me to redistribute splits around apiaries more easily than if other methods have been used. There are a few variations that can be done with different aims, but I will discuss these later.

The process is straightforward – when open queen cells are discovered on an inspection bring a nuc box to beside the colony. Remove the frames and place these within easy reach. Continue inspecting the original colony, noting any frames that have queen cells started on them. Personally, I think of the frames as having a number and do so by memory – but I have in the past put a drawing pin onto the frames will cells as an aide-memoire. Continue to inspect until the queen is found. Once she is remove her into the nuc. In a perfect world, she will be on a frame of emerging brood with lots of workers and the nuc can be refilled with frames and closed for a moment to allow the beekeeper to return their attention to the original colony. At this point it is necessary to decide what you intend to do with both halves.

Early swarm preparations might mean that both halves have a reasonable chance of building up to a point where they can still produce a crop – in which case - consider whether there are enough bees to support the queen as she rebuilds. If not, additional frames (checked to ensure the absence of queen cells) or bees shaken in can be added. This nuc can then be removed to another site and treated essentially as a queenright split. Depending on the conditions you may wish to add additional stores or to feed.

Later preparations may mean that it is unlikely that both halves will have sufficient time to build up to produce a crop and so if this is the case, the nuc can be placed in the original position and receive the foraging bees – meaning it will rebuild at a rapid rate. Or the original colony can be allowed to receive the flying bees meaning that it will continue to maintain production for a period of time. This can be useful prior to a flow of a crop such as field beans for instance.

Having made your decision, it is now important to turn your attention to the original colony and remove all but one of the queen cells. Because of the importance of nutrition in a colony being able to produce the very best queens – it is a good idea to focus on the smallest larvae in the cells, to enable the maximum feeding of this cell. It is a good idea to mark this frame – as you will want to take extra care when inspecting the following week. If this colony is being moved across to enable a nuc to collect the foraging bees, this can be done and the nuc placed in it's previous position.

The important final step is to return to the original colony and inspect it the following week. As there was a queen in the colony the preceding week, it is likely that the colony will have attempted to create additional queen cells. These will all need to be removed to prevent swarming, leaving only the cell from the first week. After this the colony can be left for a minimum of a fortnight (and ideally at least 3 weeks) before inspecting as it will not be able to produce any more cells.

CHAPTER 6:

MATING NUCS

One of the most rewarding aspects of beekeeping can be rearing your own queens. At times this can be quite a strain on resources, particularly when it comes to creating the nuclei for queens to mate from. The mini nuc is an excellent alternative that can reduce the amount of resources that are required during this stage of raising your own queens – although not without a substantial amount of additional steps required by the beekeeper. In my experience, unlike many areas of beekeeping where there are shortcuts and simpler ways to reduce workload – the setting up of mini nuclei is not one of these cases. To achieve the best results, it is important to follow the steps and timings.

Why use mini-nucs? There are several benefits, aside from the reduced number of resources required to establish them. One example is that queens will tend to mate and come into lay more rapidly in mini-nucs. There are various ideas about why this may happen – from the reduced size making it harder for virgin queens to hide from workers chasing them out of the hive to the increased stress to the small colony to have a laying queen. Regardless of the reasons behind this, this is beneficial to us in the UK due to a relatively limited season for getting queens mated.

Secondly, many methods of queen rearing produce a significant number of cells. The use of mini-nucs enables the beekeeper to use a larger number of these rather than being forced to either give them away or destroy them. A relatively small operation can easily provide the nurse bees to stock a number of mini-nucs. Additionally, it a queen fails to mate or becomes a drone layer there is a vastly reduced waste of bees in setting up.

Preparing the Nucs

There is some preparation that goes into getting the mni-nucs ready for use. All will use smaller sized frames, some constructed of wood and others injection moulded plastic. To facilitate the bees drawing comb, it is beneficial to add wax to these. Starter strips can be used, although coating the plastic frames with melted beeswax is equally effective. Following this, you have a choice of how you get the frames drawn out. Many simply fill the boxes with the nurse bees required and allow them to do the work of building the comb. This obviously has the benefit of being quicker and requiring less effort at this point. Alternatively, you can relatively easily convert a super into a rack (similar to a section rack) and place this above an established colony. Whilst this measure seems to be an additional step, the construction of the rack will become useful later in the process. The combs do not need to be fully drawn out – but are ready to use after about a third has.

The next step in preparing the mini-nucs is to provide feed. Generally, fondant is the easiest to use. The principle reason for this is that when the mini-nuc is filled the fondant will stay put. It is also absolutely fine to continue to use fondant through subsequent uses. Syrup risks spillages – although it is an option to use when reusing a mini-nuc after removing a queen. An important step here is to note the need for adding material that will float to make it easier for the bees to use the syrup without drowning. Some small chips of cork will work adequately.

Adding Bees

Mini-nucs need to be stocked with young bees. This is one of the essential steps in setting them up. Choose a donor hive that is a thriving colony. Multiple hives can be used, so as to avoid depleting any individual hive too much. Find the queen and isolate her – a crown of thorns will do, but you could remove her to a nuc box temporarily.

Removing combs with brood from the hive, give these a firm shake into a large container. A bucket or gardeners trug are good options. Be careful not to deprive the colony of too may of its young bees. Once the bees have been collected, a further shake (many beekeepers use a kick!) will cause the older bees to fly, thus removing them from the container. The bees that remain should then be given a gentle spray of a weak sugar solution.

At this point they can be scooped into the mini-nuc. Each different brand of mini-nuc will have a required amount, generally between 250 and 300ml of bees. This might seem a confusing element to many – it means you need to have measured that volume in advance. A disposable cup is an easily available measure. Prior to stocking the nucs, use a measuring jug to find the correct volume of water. Fill the disposable cup to this level and then mark it at that point. Ideally, using something that won't rub off or disappear with use (tape works quite effectively). You can then use this as your measuring container to make sure there is the needed quantity of bees. It is important not to overfill mini-nucs as they are prone to overheating and this will contribute to the problem. These are then poured into the mini-nuc which is then closed completely. The nuc will need to be left for enough time to realise that it is hopelessly queenless. It should be a minimum of a few hours, but ideally 3 days. As the mini-nuc will be sealed for this time, it important that it is placed somewhere cooler – out of direct sunlight – and that it is sprayed with water to aid in cooling and the bees accessing the fondant.

Adding a Queen or Cell

Once the mini-nuc is at this stage, it is time to add a queen cell or a virgin queen. Adding a queen cell is relatively straightforward. The essential element here is the age of the cell. It needs to be within 1 or 2 days of emergence. The difficulties that mini-nucs have in regulating their temperature mean that they are not a suitable replacement for a cell finisher or incubator. Effectively, the queen needs to be fully formed in order to successfully emerge. Installing the cell varies slightly by the type of mini-nuc used – for instance in an Apidea there is a liftable flap in the plastic 'cover board' big enough to insert the cell. They are designed with the Nicot style kit in mind, and the cell holder will then rest in between the frames. Most work in this way, however if you use JZBZ cups be prepared that they may not fit as snuggly and are prone to dropping down. There is some debate as to whether to check for emergence – some beekeepers like the reassurance of being certain, while others prefer not to risk a flighty virgin disappearing off over the horizon.

Introducing a virgin queen avoids the need for this check, and is a relatively straightforward process. As soon as possible after emergence, the queen needs to be added to the mini-nuc. Depending on where she emerged, it is a good idea to dip her in a glass of water prior to droping her into the mini-nuc. This serves to remove the scent of another colony (if she emerged in a hair roller cage within a cell finisher) and encourage the other bees to groom her aiding in her incorporation into the colony. Whilst not absolutely necessary if the queen emerged in an incubator, it does no apparent harm.

Once introduced the mini-nuc should be closed and remain this way for 24 hours. This is the ideal time to transport the mini-nuc to the mating apiary. This should be away from the main apiary site. Ideally, the mini-nucs will be placed on stands and it is well worth considering helping the queens that return from their mating flights to identify their own nuc. This can be done by painting the mini-nucs in different colours or by using simple patterns to aid distinguishing them from one another. Orientating the nucs in different directions and making sure they are well spaced out will further help. Try your best to achieve as many of these elements as you can, but sometimes only a few will be possible because of the location of the apiary.

Checking Mating and Reusing Mini-Nucs

After two weeks, and given the right conditions, the virgin queen should have mated and come into lay. This can be determined by looking for sealed worker brood. If none is present, then reassemble the mini-nuc and wait another week. However, if there is worker brood it is a good idea to remove the successfully mated queen or add additional space if planning on overwintering her in a mini-nuc.

Having removed the mated queen, it is entirely possible to reuse the mini-nuc which will now be full of bees and brood. It is important to check on stores, as these may need to be replenished. Then you can add a queen cell within a day or two of emergence. In reusing mini-nucs, queen cells are generally better tolerated than virgin queens. The same process of dipping a virgin in water can be used, but the acceptance rate seems to be lower. Generally, mini-nucs can be reused multiple times throughout a season provided that they are successful on the preceding attempt. There is very little point in attempting to reuse a failed attempt – the bees will be aging, without any replacement bees emerging from brood produced. In this case, it is better to shake out the bees and start the process again if required.

At the End of the Queen Rearing Operation

Once the number of required queens have been produced, any remaining brood need not be wasted. This is where the super/rack described earlier becomes particularly useful. This can be combined to another colony allowing the brood to emerge and join the larger colony. The combs should then be preserved – providing drawn comb for the following season.

CHAPTER 7:

NUCS FOR OTHER PURPOSES

There are other uses for nucs – adding to the benefits of always having a few nucs throughout the season.

Observation Nucs

Whilst public awareness of the honeybee has hugely increased over the last few years, it is always a pleasure to share the craft of beekeeping with those who otherwise not have the chance to get to see a hive up close. The observation nuc is the perfect opportunity to do this!

Whilst they are relatively straight forward to set up, there are some additional considerations to ensure the welfare of the bees and the ease of viewing. As I use an observation hive at various events throughout the season, I find it easiest to have one or two nucs specifically for this purpose. Part of the advanced preparation of these nucs is ensuring that the comb is relatively fresh and certainly not dark, older comb which is nearing being replaced. This will make the observation hive more visually appealing and often easier for viewers to see the range of different activities that are taking place inside. The second advance element is to manage the strength of the nuc. For observation purposes, a really strong colony is not as one that is more moderately populated. Ideally, the queen will be marked – to make it that bit easier for people to attempt to spot the queen, and if possible a queen that will calmly walk about the combs allowing people to see her rather than one of those queens that likes to hide herself away.

The next is when it comes to setting up the nuc prior to taking it to an event. The first step is to find the queen and to remove her to the upper box where she will be on display. Pay close attention to the frame she is on – ideally there will be a good balance of stores, pollen, sealed brood, open brood, eggs and space to allow the queen to continue to lay. This will give anyone looking plenty to see and ask questions about. It is particularly important to try to avoid frames that are largely eggs – as these require relatively little attention from the workers and this will leave the frame on display less interesting for people to see. Once this frame is in and sealed in the box, it is a good idea to try to bleed of some of the flying bees as they can obscure the view and seem to run around inside an observation hive as they try to go about their work. This can be done by giving each frame a tap to remove foragers before it is placed into the lower chamber. Importantly, ensure that the colony has plenty of stores – ideally open nectar for ease of access. They can be provided within an internal frame feeder but it is vital that this has straw or corks added to ensure that the bees do not drown. When it come to sealing the nuc, most observation hive will have clasps to keep them closed. If being shown to the general public, particularly if children are present, it is a good idea to ensure that these clasps are secured – depending on the number or location, cable ties or padlocks might be appropriate. An alternative it to use the clasps as intended, but secure the sections together using screws. Once the nuc has been assembled, try to keep it cool and if not using an internal feeder make sure that the colony has supply of moisture. A spray under the mesh floor every so often will help. Ideally, the bees should not be kept in the observation hive for too long - over a weekend event is possible when they are completely sealed inside but is about the limit before having a significant impact on the colony.

A slight step up from an observation hive is a demonstration nuc – ideal for showing those that are more interested in beekeeping and would like to have a more detailed view, such as a beginner or those looking for beekeeping experience. These can be managed in such a way as to maximise the potential for this to be an enjoyable experience and minimise the chances of overwhelming those with limited experience by large volumes of bees or with being stung. The same principles could be used to prepare a full sized colony for the same purpose.

In an ideal world, every colony that beekeepers kept would be calm and not likely to become defensive during a prolonged inspection. The reality is that circumstances often mean that we have to work with bees when this is not the case. However, by considering honey bee biology it is still possible to prepare colonies for demonstration purposes that can be a pleasure to inspect. The key factor is age polyethism – that workers progress through a range of different tasks during their lifetimes. This means that with some manipulations it is possible to remove older bees and thereby reduce the number of bees and their ability to deliver stings.

In order to do this, it is very similar to performing a split. Ideally, a strong nuc will be used as it will lose a substantial proportion of it's work force. The nuc to be demonstrated needs to be moved to a different location in the apiary at a time when the foragers are flying. This means when they leave the hive they will return to the original location. A weaker colony or one in the process of building up can be placed in the original location so as to benefit from the additional workers. This will leave the nuc for demonstration largely compose of younger bees. At this point it can be provided with any additional resources it needs (for instance a frame of open nectar, a frame of pollen). This can then be transported to where it is required for the demonstration. It is a strong suspicion that most of the social media content of beekeepers inspecting colonies without protective equipment use this technique as the photo below demonstrates.

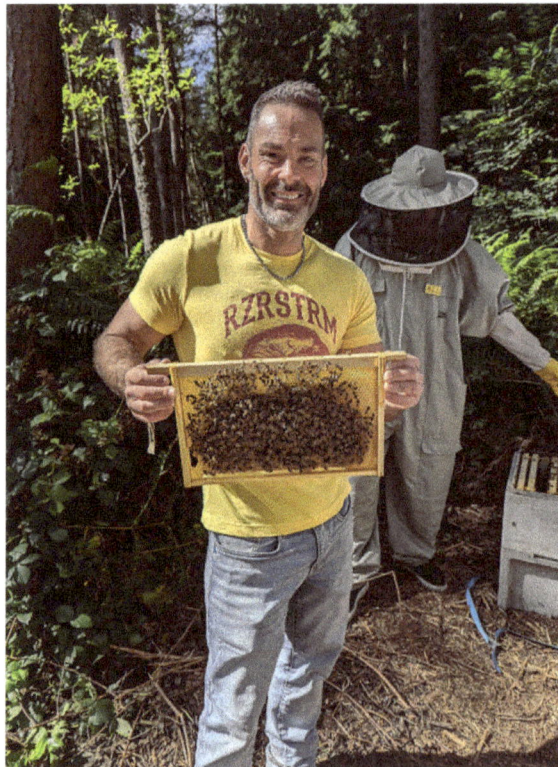